Sensors for Sustainability 2022

Topical Conference at the 2022 AIChE Annual Meeting

Phoenix, Arizona, USA
13-18 November 2022

ISBN: 978-1-7138-7911-4

Printed from e-media with permission by:

Curran Associates, Inc.
57 Morehouse Lane
Red Hook, NY 12571

Some format issues inherent in the e-media version may also appear in this print version.

Printed with permission by Curran Associates, Inc. (2023)

For permission requests, please contact AIChE
at the address below.

AIChE
120 Wall Street, FL 23
New York, NY 10005-4020

Phone: (800) 242-4363
Fax: (203) 775-5177

www.aiche.org

Additional copies of this publication are available from:

Curran Associates, Inc.
57 Morehouse Lane
Red Hook, NY 12571 USA
Phone: 845-758-0400
Fax: 845-758-2633
Email: curran@proceedings.com
Web: www.proceedings.com

TABLE OF CONTENTS

SENSORS AND MONITORING FOR HEALTH

GENERAL POSTER SESSION IN SENSORS

FIELD-DEPLOYABLE SENSORS

NEXT-GEN SENSORS

STUDENT COMPETITION IN SENSORS (SPONSORED)

TOPICAL PLENARY: ENVIRONMENTAL ANALYTICAL TECHNOLOGY: SENSOR APPLICATIONS IN SUSTAINABILITY (INVITED TALKS)

92c Fabric-Based Wearable Health Sensors

Chelsea Monty-Bromer

The need for well-managed hydration and the monitoring of physiological parameters in real-time is not only crucial for athletes to optimize performance but to also prevent illness and injury from occurring. Understanding sweat loss and calculating hydration requirements to avoid dehydration or overhydration is a significant challenge for athletes. This is because sweat losses vary significantly from athlete to athlete both in terms of concentration of electrolytes and total volume of sweat lost. When sweating, an individual may lose as little as ~200mg of sodium per liter or as much as ~2000 mg. This variance, combined with significant differences in sweat volume, lead to vastly different hydration requirements among individuals. The RooSense SweatID system is the first lightweight fabric sensor to provide real-time information regarding hydration levels during exercise or training through selective determination of sodium levels, a better marker for athletic performance, safety, and injury.

Monitoring sweat is a compelling choice to gain insight into a person's health at the molecular level, as sweat is rich in physiological and metabolic information that can be obtained non-invasively. In spite of advances made towards the detection of biomarkers in sweat, there is no sensor capable of long-term detection in constricted or load-bearing applications where other flexible plastic sensors might cause discomfort. Textile-based and nanocomposite technologies provide an optimum scaffold for monitoring biomarkers at the surface of the skin as they do not influence the sweat composition or sweat rate and can be seamlessly integrated for wearable sensing and translational medicine applications.

This presentation will focus on the development of a prototype wearable hydration monitor, validation of the prototype in the laboratory, and on body testing in an exercise physiology laboratory.

92d Dopamine Modified Carbon Fiber Micro-Electrodes for Enhanced Detection of Cu(II) Via Fast Scan Cyclic Voltammetry

Noel Manring, Muzammil Nishar Ahmed, Jonathan Xavier, Lexine Sibert, Pavithra Pathirathna

Neurological disorders (NDs) affect millions of people globally despite the meticulous efforts of the medical community to limit their progression. The lack of efficient drugs to address NDs underlines the need for a deeper understanding of the etiology of these diseases, to develop effective medications. One such aspect is the effect of external factors such as heavy metals on the progression of NDs. Cu(II) when present in excess is known to play a pivotal role in NDs such as Alzheimer's disease. Copper's ability to cycle between oxidation states when bound to amyloid-β aggregates leads to an overproduction of reactive oxidative species, causing irreversible oxidative damage. Therefore, it is important to monitor the levels of Cu(II) in neurological systems to increase the understanding of its impact on NDs.

Many analytical tools have been developed to detect trace amounts of Cu(II). However, conventional methods such as spectrometry, chromatography, immunoassays, and colorimetry analyze blood or urine samples or postmortem tissues. These methods are not translatable to living organisms and limit access to real-time information. Thus, there is a critical need to develop a sensor with a high temporal and spatial resolution to study the changes of heavy metals such as Cu(II) in living systems in real-time. In recent years, electrochemists have made a breakthrough in developing sensors targeted toward detecting the real-time concentration of toxic metals. However, most of the reported electrode fabrication used to enhance the sensitivity are not bio-compatible, making them a poor choice for in vivo measurements. Therefore, the development of a cheap and bio-compatible surface modification for detecting physiological concentrations of metals ions in vivo will largely aid in bridging the knowledge gap in understanding the vital roles heavy metals play in NDs.

This study uses fast-scan cyclic voltammetry (FSCV) with dopamine-modified carbon fiber microelectrodes (CFMs) for the detection of trace amounts of Cu(II). Mussel-inspired chemistry has shown that dopamine can be used as a surface modification by mimicking the strong adhering abilities of the marine mussel. The CFM was fabricated by the electrodeposition of dopamine, making it a bio-compatible surface modification that is ideal for in vivo measurements. All the experiments were performed in tris buffer, which mimics cerebellum fluid. We optimized the electrodeposition process by characterizing the electrochemistry of dopamine with our sensor and selecting the best time for electrodeposition based on FSCV experimental results and the surface morphology of the electrodes observed using atomic force microscopy. We characterized the sensor for the detection of Cu(II) with and without surface modification and find that there is a significant increase in the sensitivity of the dopamine-modified CFM. We also performed stability and shelf-life tests to examine the analytical performance of our sensor for prolonged usage.

The bio-compatible nature of the dopamine surface modification coupled with enhanced sensitivity and the high temporal and spatial resolution provided by FSCV allows us to make fast in vivo measurements and promises great potential for further development into a powerful analytical tool for the detection of Cu(II) in neurological systems. Future studies will focus on extending this research toward other toxic metals and developing a multichannel system for the simultaneous detection of neurotransmitters and toxic metals.

92e Highly Selective Biosensor Based on Ionic Liquids-Assisted Colorimetric Sensing for Arsenite

Sam Li, Xuan Hao Lin

We present a highly selective biosensor based on colorimetric detection with ionic liquids (ILs). The agglomeration of gold nanoparticles (AuNPs) caused by ionic liquids was prevented by arsenite (As(III)). This study looked at six distinct ILs, including 1-(3-cyanopropyl)-3-methylimidazolium chloride (CPMI), 1-(3-cyanopropyl)pyridinium chloride (CPP), tetradecyldimethylbenzylammonium chloride (TDDBA), cetylpyridinium chloride monohydrate (CP), 1-butyl-1-methylpyrrolidinium methyl carbonate (BMP), and (TBMP). We discovered that a series of ILs were up to 40,000 times more efficient than standard inorganic electrolytes (e.g. NaCl) in producing AuNP aggregation, which was blocked by As (III). Molecular geometry optimization and density functional theory (DFT) calculations were used to investigate two- and three-way non-covalent interactions between ILs, citrate, and metal ions. Arsenite may be distinguished from arsenate and a variety of other ions such as cadmium, zinc, sodium, chromate, bromide, and iodide by inhibiting ILs-induced AuNPs aggregation. The findings of DFT calculations explain why BMP were only binding to arsenite. Repeatability, interference, stability, selectivity, and sensitivity were studied. Detection limits were further lowered by a random DNAzyme substrate. The limit of detection (LOD) of the sensors was 10 ppb. The arsenite sensing was demonstrated using real water samples and validated with inductively coupled plasma-optical emission spectroscopy (ICP-OES).

92f Dynamic Measurement of Endogenous Acetone in Exhaled Breath As a Non-Invasive Alternative to Obtain Diffusing Capacity of the Lung

Anastasios Angelopoulos, Jonathan A. Bernstein

Diffusing Capacity of the Lung for Carbon Monoxide (DLCO) is a well-established procedure for measuring the efficiency of pulmonary gas exchange to diagnose pulmonary diseases and particularly Idiopathic pulmonary fibrosis. The method uses a tracer gas such a helium to obtain lung volume and an exogeneous source of carbon monoxide to provide a well-defined concentration gradient. Ultimately, one obtains the mass transfer coefficient of alveolar lung tissue and associated resistances. In this work, we discuss development of a sensor that uses of endogenous acetone as a non-invasive and more convenient alternative to DLCO.

Given that a typical breath exhalation can be sustained no longer than ~40 s, any exhaled breath analysis sensing method requires time resolution on the order of 1 s to be able to deconvolute the time-scales of the mass transport processes involved. Existing portable detection methods for endogenous volatile organic compounds (VOCs) such as acetone rarely achieve such time resolution and rely on ex-situ analysis of a collected breath sample. Furthermore, water in exhaled breath presents a significant interference for typical potentiometric or optical sensing approaches. We have previously shown that immobilized organic reagents in perflurosulfonic acid membranes can undergo a rapid water-resistant color change by reacting with acetone in exhaled breath. Our recent investigations have disclosed how such behavior is enabled by differing transport pathways in the membrane for acetone and water. We will here present application of this optical sensing approach in small scale clinical investigations that reveal the alveolar mass transport resistance to acetone transport. Comparison to DLCO results will be discussed.

375a Pd-Nanoparticle Enabled Optical Fiber Hydrogen Sensor for Subsurface Storage Conditions

Daejin Kim, Nathan Diemler, Ruishu Wright, Michael Buric, Paul Ohodnicki

Hydrogen is the center of attention as the hydrogen economy is expected to aid significantly in the decarbonization of the energy and transportation sectors. As more hydrogen is produced and utilized to decarbonize, more storage for surplus hydrogen will be required. One of the cost-effective methods to store hydrogen at large scale is to use underground resources such as salt caverns, aquifers, and depleted oil and gas fields. In those underground storage systems, it has been found that hydrogen concentration is subject to change due to microbial reactions and leakage through caprocks and wells. Therefore, it is critical to monitor hydrogen concentration in a real-time manner under humid and harsh subsurface conditions. We have developed an optical fiber sensor coated with palladium nanoparticle incorporated-silica composite layers to detect hydrogen over a wide range of hydrogen concentrations. The effects of palladium particle-loading and thickness of the composite layers on hydrogen sensing performance with optical fiber sensors will be discussed in the presentation. The hydrogen sensing response under different humidity and temperature conditions that simulate the subsurface hydrogen storage conditions will also be presented. The optical fiber sensor developed in this study would enable real-time hydrogen monitoring in the subsurface hydrogen storage systems in a safe and stable mode.

375b Reshaping Molecularly Imprinted Polymers for Robust Sensing Performance

Cameron Malloy, Suchol Savagatrup, Ramon Castrejon Miranda

Molecularly imprinted polymers (MIPs) function as synthetic analogues to antibody-antigen systems with a "lock and key" mechanism for molecular recognition. In combination with an electrochemical interface, MIPs have been used to detect a diverse range of biological and chemical analytes, e.g., carcinoma biomarkers and perfluoroalkyl substances (PFAS). However, current challenges exist with MIP sensors specific to detection reliability, film reversibility, repeated reusability, and transport phenomena dynamics. These intrinsic challenges limit the MIP sensor robustness and practicality in real word scenarios. In this work, we present a materials science approach to address the current MIP shortcomings through tuning of the copolymer film composition. Varying the copolymer film makeup through electrophilic aromatic substitutions of di-substituted phenylenediamines has shown enhanced MIP performance and reliability as determined through potentiodynamic analysis. Specifically, we target cross-link density to affect the degree of swelling and contraction during the solvent extraction process to improve film reversibility. Furthermore, we then control film porosity to enhance non-covalent interactions during anodic electrochemical polymerization and measure the impact on adsorption and desorption kinetics for the target compound. We anticipate this materials science approach will address current limitations with MIP sensors and provide a framework for robust MIP sensing performance.

375c Dynamic Complex Emulsion Based Multiplexed Sensing Array for Environmental Contaminants Detection

Baishali Barua, Tyler Durkin, Suchol Savagatrup

Continuous and real-time monitoring of emerging environmental contaminants is essential for acquiring actionable information and detecting subtle changes of various transients in a real-world environment. Many laboratory-based, analytical techniques for the detection of contaminants (*e.g.*, heavy metals and per- and poly-fluoroalkyl substances, PFASs) are cost-prohibitive for large-scale, distributed sensors. Moreover, they are often not suitable for rapid, on-site, and frequent detection. Hence, we report a novel sensing platform that utilizes multicomponent emissive complex droplets as multiplexed sensing elements to detect multiple environmental contaminants simultaneously. Specifically, polydimethylsiloxane (PDMS) microfluidic chip is fabricated as a sensing chamber that holds complex droplets and facilitates continuous monitoring of targeted analytes through an optical detection method. We measure the dynamic changes in interfacial tensions at multiple interfaces of fluorophore embedded complex droplets while interacting with the targeted chemical compounds. These measurements are correlated with the concentrations of targeted chemical compounds. Additionally, we test the selectivity of the sensing platform towards the analytes in a realistic matrix of synthetic groundwater. This framework promotes a unique and robust method of sensing that can provide rapid, on-site detection of environmental contaminants in complex matrices.

375d Stress Sensor Synthesis Using Electropolymerized Molecularly Imprinted Polymers

Grace Dykstra, Yixin Liu

Molecularly imprinted polymers (MIPs) are synthetic receptors that mimic the 'lock and key' mechanism in the natural biological antigen-antibody system. They are highly attractive as artificial receptors with tailored biomolecular recognition to construct biosensors. Electropolymerization is a quick and easy method to directly synthesize MIP sensing elements *in-situ* on the working electrode, enabling ultralow-cost and easy-to-manufacture electrochemical biosensors. Finding an optimal design space and relationship between synthesis parameters and sensitivity is one of the most challenging aspects of MIP fabrication. e -MIPs' development and optimization can be facilitated through the use of machine learning to quickly identify this design space and improve performance. For cortisol detection, pyrrole is selected as the functional monomer. Cortisol-imprinted-polypyrrole-based sensors are fabricated with 72 sets of synthesis parameters with replicates for the data-driven analysis, including pyrrole concentration, cortisol concentration, number of electropolymerization cyclic-voltammetry (CV) cycles, electropolymerization CV scan rate, and number overoxidation cycles for template elution. MIP sensing performances are measured using a 12-channel potentiostat, which improves experimental efficiency and ensures sensor reproducibility and data quality. Machine learning optimization can also identify coupled parameters more efficiently than a traditional design of experiments. These linkages assist in the development of enhanced sensors for cortisol and the method can be expanded and applied to other MIP sensor models.

375e Enzymatic Electrochemical Sensing for Fish Freshness Using Macro-Porous NiO Electrodes

Anuja Tripathi, Anastasia L. Elias, Abebaw B. Jemere, Kenneth D. Harris

Introduction

The freshness of fish meat is the most important criterion in the pre-consumption quality control process. Complex microbiological processes lead to the loss of freshness and subsequent spoilage. Six hours after death, autolytic decomposition of adenosine triphosphate (ATP, the energy-storing molecule found in cells) begins to produce xanthine (XA) as one of the primary metabolites. XA concentration increases with storage time, and thus it can be used as an indicator of fish freshness.

Current XA determination methods, which use chromatography, colorimetry, spectrophotometry, or mass spectrometry, are time-consuming, expensive, and require skilled operators. Electrochemical sensor platforms, on the other hand, are attractive because of their ease of use, low cost, small sample volume requirements, and quick analysis. In recent years, enzyme-based electrochemical biosensors, employing xanthine oxidase (XO) for selective recognition and catalysis of XA, have become promising for XA determination in fish and meat samples because of their low detection limits, high selectivity, and sensitivity.

The working electrode material has a significant impact on the analytical performance of electrochemical biosensors, and the integration of nanomaterials in the working electrode has previously improved sensor performance. Although both carbon-based and metal/metal oxide nanostructures with favorable electrical characteristics and biocompatibility have been reported for enzymatic electrochemical biosensor fabrication, macroporous NiO electrodes are rarely employed for enzymatic detection of XA.

Macroporous NiO electrodes are ideal for electrochemical biosensor fabrication owing to their large surface area for enzyme immobilization, high isoelectric point, biocompatibility, chemical stability and fast electron transfer features. Furthermore, NiO possesses its own redox pair (Ni^{3+}/Ni^{2+}), allowing for reagentless biosensing. Here, we describe macroporous NiO electrodes fabricated by glancing angle deposition (GLAD). GLAD is a physical vapor deposition technique capable of producing meso- and macro porous thin films with extremely high surface area and internal porosity. The large surface area and porosity of the GLAD films should allow immobilization of large quantities of the XO enzyme, which, when combined with the excellent electrical properties of the GLAD electrode, is expected to allow for the amperometric detection of XA. In this work, XO is immobilized on a NiO GLAD electrode, and the resulting sensor is used to determine the concentration of XA in a fish sample with excellent sensitivity, detection limit, and response time.

Sensor fabrication and performance

Nanocolumnar and mesoporous NiO films (thickness ~500 nm) were deposited on ITO using GLAD. XO was physisorbed onto electrochemically activated GLAD NiO electrode by incubating in XO-loaded PBS buffer, pH 7.4 at room temperature. The modification of the NiO GLAD film with XO was characterized by electrochemistry and X-ray photoelectron spectroscopy. XO catalyzes the oxidation of XA to uric acid and hydrogen peroxide (H_2O_2), and therefore, using an optimized measurement condition, the current was monitored by applying +0.5 V vs Ag/AgCl to the electrode. This current was found to be proportional to the concentration of XA. The amperometric sensor rendered a dynamic range of 0.1 μM to 650 μM, with a limit of detection of 37 nM, good reproducibility (relative standard deviation of ~ 4%, n=18), rapid response time (~7 s), and a high sensitivity (1.1 $\mu A \cdot \mu M^{-1} \cdot cm^{-2}$ in the low concentration range from 0.1-5 μM, and 0.3 $\mu A \cdot \mu M^{-1} \cdot cm^{-2}$ in the higher concentration range from 5-650 μM) compared to the flat NiO-based enzymatic sensor (0.03 $\mu A \cdot \mu M^{-1} \cdot cm^{-2}$). The sensor showed little interference from common fish sample matrices (such as glucose, uric acid, hypoxanthine) and fish preservative (sodium benzoate), and was used to quantify XA in real fish samples. The sensor could also be stored in buffer at 4°C for over a week without losing its performance.

Overall, the developed enzymatic electrochemical biosensor exhibited good performance to monitor fish spoilage. We believe this study will contribute to a better understanding of macro-porous NiO electrodes, which will aid in the development of electrochemical biosensors for other compounds.

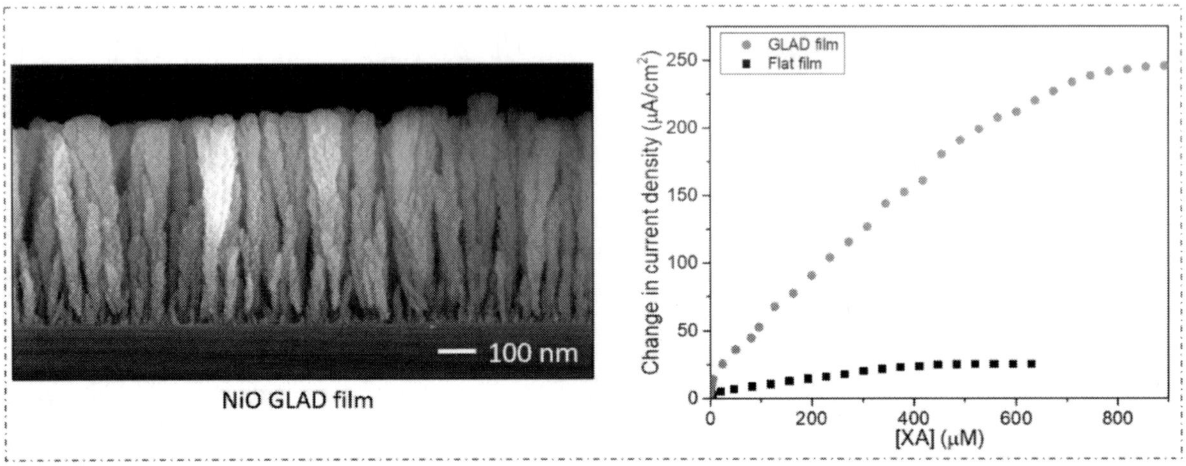

287a Invited Talk: Field Deployable Sensor Arrays with Molecular Selectivity

Thomas Thundat

Microfabricated sensor array is an ideal sensor platform for field deployable miniature sensors. Many such platforms, such as cantilever arrays, have been demonstrated as sensitive sensors for chemical and biological detection. In fact, the array-based sensor platform satisfies almost all of the requirements of an ideal high-performance sensor, such as its very high sensitivity, miniature size, array based multi-analyte detection, and low power consumption. However, despite all of these advantages, many of these sensors still remain only a laboratory tool without commercial success. Poor molecular selectivity, especially for small molecules, as well as the lack of reproducibility, pose formidable challenges in translating this highly sensitive sensor platform into a commercial reality. Enhancing selectivity and sensitivity of these sensors has been the focus of intense investigations over the last decade or so. Traditionally, selectivity in chemical and biological sensing is achieved by using immobilized receptors or chemical interfaces on sensor surfaces. While poor selectivity in small molecule detection using reversible receptors is due to the lack of uniqueness of receptor-analyte interaction, nonuniformity in the immobilized receptor graft density is the reason for unacceptable rates of reproducibility in these sensors. Various strategies are currently used for overcoming the selectivity challenges. These approaches include separation prior to detection and incorporation of multi-modal, multi-physics approaches. Multi-modal, multi-physics data obtained with these approaches, when analyzed using deep learning techniques, show enhanced selectivity, sensitivity, and reliability. I will discuss recent advances as well as the challenges in the development of field deployable miniature sensor arrays for detection of small molecules in vapor phase.

287c Novel Field Deployable Sensors to Monitor How Dynamic Hydrology Shapes Nutrient and Element Transformations in a Great Lakes Coastal Estuary

Chelsea Monty-Bromer, Sai Prasanna Chinthala, Bukola Adesanmi, Zachary Cheney, Lauren Kinsman-Costello, John Senko, Joshua Davis

The propensity for redox processes to occur is often predicted using measured or assumed redox potential (E_h). However, process measurements often deviate from predictions based on the traditional, thermodynamic "redox tower" paradigm. In this work, we hypothesize that poorly measured, granular scale spatial and temporal soil heterogeneity causes apparent departures from thermodynamic exclusion principles at bulk scales, with consequences for large-scale biogeochemical cycling which are not currently resolved in ecosystem, regional, or global scale models.

We developed a 3D printed graphite electrode in order to perform zero resistance ammetry (ZRA) measurements to measure redox disequilibria at various spatial scales in benthic sediments. This allowed us to detect the distributions, extents, and kinetics of biogeochemical processes. In this work, ZRA was used to measure electrical current that arose from microbially-induced redox disequilibrium. We compared the 3D printed electrodes to graphite mesh electrodes using cyclic voltammetry. Prior to deployment in the field, we used the 3D printed graphite electrodes in opposing chambers of a split cell setup containing poorly crystalline Fe(III) (hydr)oxide-rich soils. When both chambers were incubated under oxic or anoxic conditions, no current was observed. However, when one chamber was oxic and the other anoxic, no Fe(III) reduction was observed in the oxic chamber and Fe(II) accumulated in the anoxic chamber. The redox disequilibrium generated between chambers was accompanied by electron transfer from the anoxic to the oxic chamber detected as ZRA.

We then refined the design, construction, and deployment of a paired ZRA and Eh multi-sensor system that will detect electrochemical signals across and directly beneath the sediment-water interface at nested scales (sub-millimeter to decimeter) at a shallow location (< 10 cm surface water) in the Old Woman Creek wetland. In summer 2022, we deployed these sensors and collected concurrent data on dissolved oxygen dynamics, surface and pore water nutrient concentrations, greenhouse gas fluxes (chamber measurements), and soil geochemistry. Future work will use this data to demonstrate the feasibility of integrating microsite electrochemical and redox variability to the ecosys ecosystem scale model for improved representation of soil redox processes in spatially variable and temporally fluctuating systems.

Specifically, our objectives are to (1) comprehensively relate dynamic hydrology to redox regimes; (2) determine how redox heterogeneity in space and time drives elemental cycling at multiple scales, and; (3) assess the sensitivity of process-based models to the inclusion of fine-scale variability in redox conditions.

287d A Silica Nanofiber-Based Colorimetric Sensor for the Point-of-Use Detection of Polycyclic Aromatic Hydrocarbons in Water

Colton Duprey, Hadi Rouhi, Hannah Stumpft, Emily Linn, Sarah Veres, George Chen, Mark Elliott, Evan Wujcik

Polycyclic aromatic hydrocarbons, or PAHs, are a class of non-polar organic compounds which are comprised of two or more condensed benzene rings. They are carcinogenic compounds that threaten long term health and in large doses, can cause acute toxicity. Since they are partially soluble in water, these pollutants that are a result of oil spills and industrial pollution are extremely detrimental to our planet's aquatic ecosystems. Herein we demonstrate the first nanofibrous polymer-based on-site colorimetric sensor for the detection of PAHs at environmentally relevant concentrations. Electrospun silica fibers serve as a reaction substrate, a support for the colorimetric reaction to occur, and to provide hydrophobicity as means to increase the concentration of PAHS out of water. The volatile PAHs are concentrated in the vapor phase via a shift in the vapor-liquid equilibrium, allowing for a more sensitive detection range than in the liquid phase. Under acidic conditions, formaldehyde is used to colorimetrically detect the PAHs—reacting with them to form brown quinoid compounds on top of the white silica nanofiber mat. Image analysis is further used to quantify the PAH concentration based off the intensity of the color change. This robust and portable sensor is an uncomplicated alternative to the traditional, expensive analytical equipment (HPLC, GC-MS), and allows for on-site and rapid water monitoring, and is the only such method of its kind to detect and concentrate PAHs out of liquid.

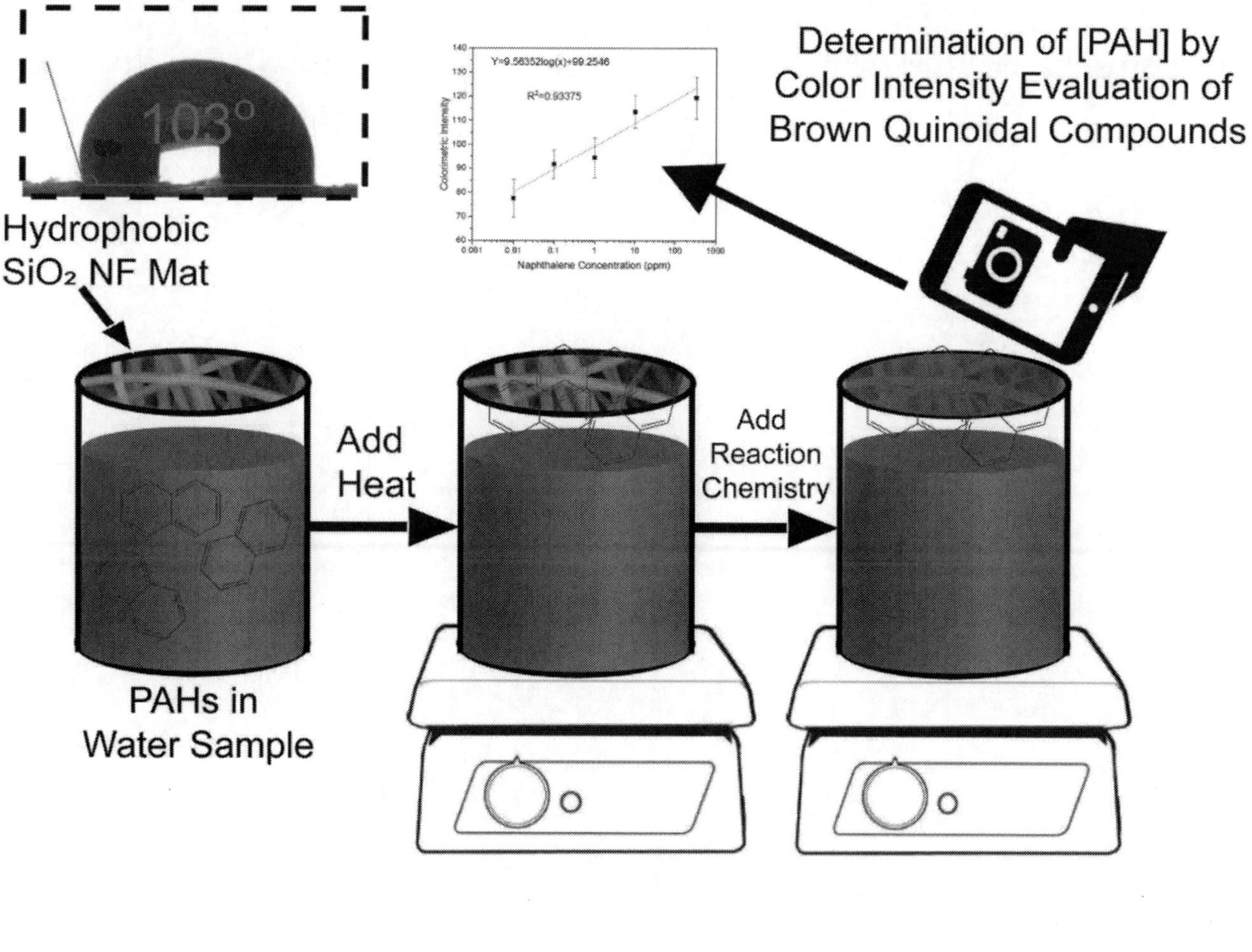

287e Point-of-Need Quantitative Detection of Trihalomethanes in Environmental Water Samples Using a Highly Sensitive and Selective Fiber-Based Preconcentration System

Hadi Rouhi, Colton Duprey, Nicole Penners, Emily Linn, Sarah Veres, George Chen, Ali Alshaikh, Yang Lu, Leigh Terry, Mark Elliott, Evan Wujcik

Nowadays with developing the industries all over the world and consequently contamination caused by their activities, whole of our ecosystem is in a threat. One of the main serious problems that we face is the by-products of water treatment system, introducing the halogenated compounds, specifically trihalomethanes (THMs), to the drinking water. US Environmental Protection Agency (USEPA) limited the concentration of THMs in the drinking water to below 80 ppb due to serious health problem these compounds occur in human body. In this study, a super hydrophobic polypropylene (sPP) fiber mat is used in a THM detection system based on colorimetric reaction chemistry. The reaction chemistry—Fujiwara reaction—has been known for over a century to give a visible red/pink chromophore under visible light absorption but has not been used for detection of halogenated species in water down to ppb-levels due to its hinderance by water. The highly sensitive and selective sensor operates when the preconcentrated THM in the vapor phase becomes trapped in the liquid phase of the Fujiwara chemistry, on the surface of the sPP fibers. sPP fiber mat is electrospun with a special "bead-on-a-string" morphology leading to a hydrophobic mat with over 150° water contact angle which ease the detection of THMs in lower concentrations in the range of ppb. The color change happened in the surface of the fiber mat can be analyzed with any image analytical software to measure the quantitative color intensity corresponded to accurate concentration of THM. In this study, we measured THM content in the water in the level of below 8 ppb with a very good approximation. This cost-effective, selective, portable device with sensitive THM detection enables on-site users to evaluate the quality of drinking water.

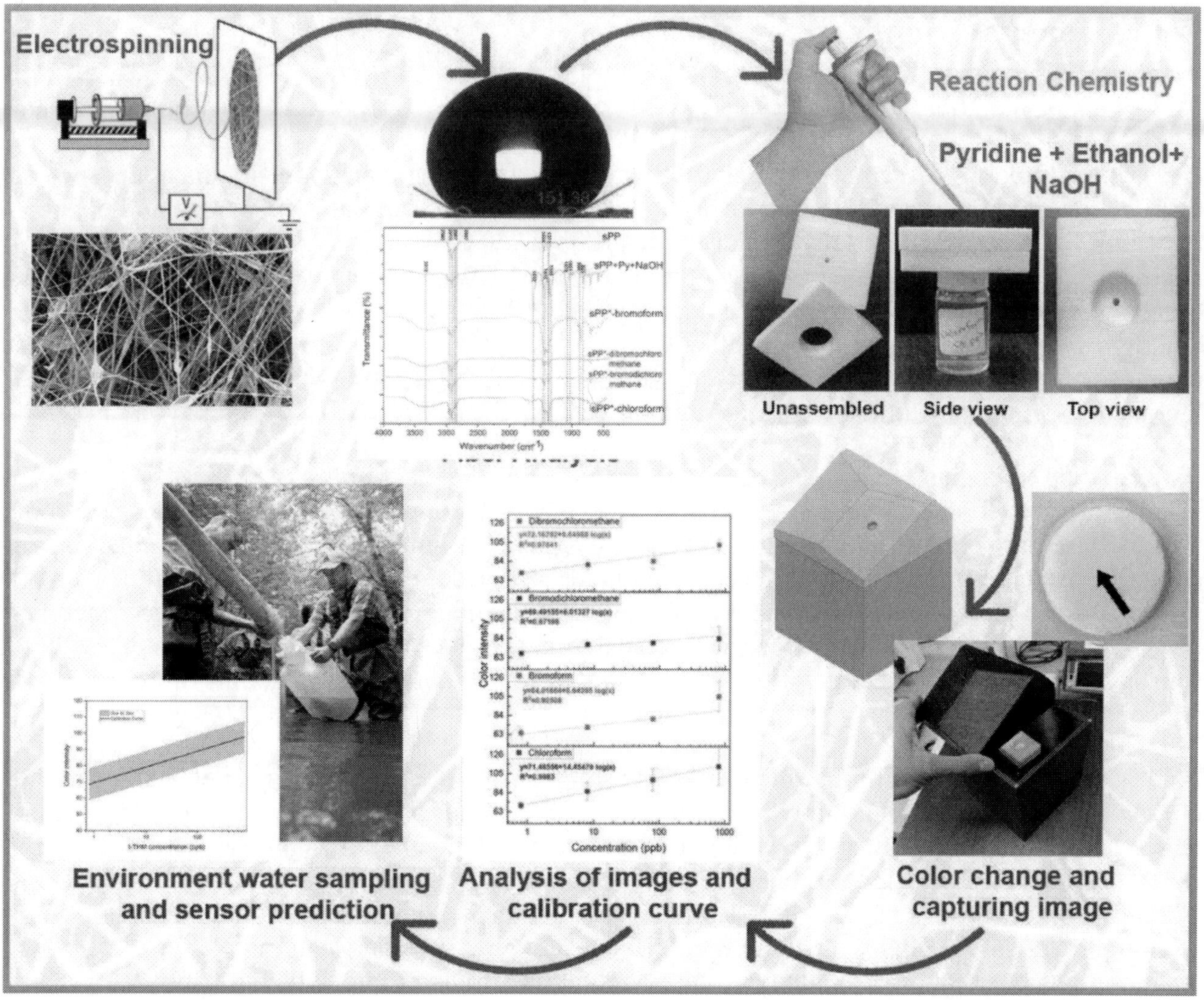
Electrospinning
Reaction Chemistry
Pyridine + Ethanol+ NaOH
Unassembled
Side view
Top view
Environment water sampling and sensor prediction
Analysis of images and calibration curve
Color change and capturing image
Concentration (ppb)
Color intensity
Dibromochloromethane
Bromodichloromethane
Bromoform
Chloroform
Wavenumber (cm⁻¹)
Transmittance (%)
sPP
sPP+Py+NaOH
sPP*-bromoform
sPP*-chloroform

287f Electrochemical Detection of Cd(II) Using Nano-Electrodes in Environmental Samples

Muzammil Nishar Ahmed, Wendy Zhou, Pavithra Pathirathna

Heavy metal toxicity is a continual global concern, which leads to deadly health disorders such as dysfunction of kidneys and livers and neurological disorders. Rapid industrialization and poor recycling methods of industrially produced metals have left humans highly vulnerable to their exposure as they bio-accumulate through the food chain. The most common route of exposure is from the water bodies where industrial effluents are dumped. The early stages of metal toxicity have common symptoms that overlap with other common diseases and thus are often misdiagnosed or not diagnosed until later. However, proper early diagnosis will largely aid in developing efficient therapeutics to treat metal toxicity, thus saving lives.

Extensive research has been done to develop various tools to detect trace amounts of metal in aqueous samples. However, these traditional techniques, such as inductive coupled plasma mass spectrometry (ICPMS), require equipment that is bulky, expensive, and difficult to operate, making them a poor choice in resource-limited settings. These techniques also need tedious sample pretreatment steps that limit real-time detection. Further, pretreatment often alters the metal speciation, a critical factor determining metal toxicity. Therefore, it is crucial to develop a robust, reliable, portable, and low-cost sensor that can provide accurate information on metal speciation of trace amounts of heavy metals in aqueous solutions.

Ion transfer between two immiscible electrolyte solutions (ITIES) is an electrochemical technique that utilizes the simple principle of ion transfer between an aqueous and organic solution. This technique is less complicated than other conventional electrochemical techniques as it does not involve any redox reactions in the system. In this study, we develop a nano-scale glass sensor that can detect Cd(II) in complicated matrices. Our electrode is a borosilicate glass electrode that is pulled using a carbon dioxide laser puller with an inner radius of ~300 nm. The nano-interface of the electrodes enables hemispherical diffusion that provides a high mass transfer rate, which is essential for fast kinetic measurements. An ionophore- 1-10 phenanthroline was used to facilitate the Cd(II) transfer across the nano-interface. The sensor was calibrated in various matrices, such as artificial seawater and tris chloride, to show its capability to withstand complicated matrices without fouling. Selectivity and stability tests were also performed to examine the sensor's performance. The sensor was also tested in the presence of different Cd(II)-ligand complexes and with a water sample collected from the Indian River Lagoon in Melbourne, FL. There's a close agreement between the concentration of Cd(II) in the lagoon determined from our sensor with previously reported values using ICPMS, thus showcasing our sensor's great potential as an environmental sensor. To the best of our knowledge, this is the first time reporting a nanometer-scale glass electrode using ITIES to detect Cd(II) in complex matrices. Further studies will be focused on detecting multiple metal ions simultaneously and developing into a portable point of care diagnostic tool to detect ingested metals in urine and blood samples.

287h Facile Carbon Electrode Modification for Improved Biosensing

Gang Fan

Modification of electrodes with biomolecules is essential for biosensor development. Though gold electrodes are often used because of their ease of modification, carbon screen-printed electrodes (SPEs) are gaining popularity, as they do not require precious metals and are more consistent than equivalent gold screen-printed electrodes. However, their modification with biomolecules remains a challenge. By combining electrochemical modification with a bioorthogonal oxidative coupling reaction, we can attach key biomolecules (DNA and proteins) to carbon electrodes. The biomolecules maintain activity following attachment, ensuring that the molecules remain active and/or accessible following coupling. We have further generated films with non-adhesive cells on these electrodes using DNA-directed immobilization. This work provides an important platform for the modification of carbon SPEs and is anticipated to be applicable to any carbon-based electrode material.

223a Invited Talk: Fabricating Ordered Nanoparticle Arrays for Biosensing Platforms Via Scanning Electrochemical Cell Microscopy

Md Maksudur Rahman, Jeffrey Halpern, Caleb Hill

Metal nanoparticles (NPs) are attractive platforms for a variety of sensing applications, but their performance is often limited by the techniques employed to immobilize NPs at interfaces. Common methods yield randomly distributed NPs which exhibit nonideal behavior due to aggregation. Well-ordered NP arrays would not suffer from these limitations, but the fabrication of such arrays is generally challenging. Here, it is demonstrated that Scanning Electrochemical Cell Microscopy (SECCM) can be used as a powerful, instantly reconfigurable tool for the fabrication of ordered NP arrays. In this approach, metal NP arrays are straightforwardly fabricated via electrodeposition at the interface between a substrate electrode and a small (sub-μm) electrolyte-filled pipet. Experimental results will be presented which demonstrate this approach enables the fabrication of well-defined Au and Ag NP arrays with spatial control over NP location comparable with the terminal diameter of the pipet employed and straightforward control over the volume of each NP. Finite element simulations of mass transport within the SECCM geometry were utilized to explain the growth kinetics and NP morphology observed experimentally. Together, these results demonstrate probe-based electrochemical techniques such as SECCM can serve as effective tools for the fabrication of NP arrays to serve as platforms for biosensing applications.

223b Anomaly Detection Algorithms to Utilize Non-Injective Gas Sensor Arrays

Paul Morris, Cory Simon

Gas sensor arrays provide an information-rich response pattern to discriminate among many gas compositions. However, the mapping of gas compositions to sensor array response vectors is often non-injective (not one-to-one), preventing their application for quantitative gas sensing tasks. Here, we show non-injective gas sensor arrays may still be useful in an anomaly detection setting.
We demonstrate computationally by (i) considering a sensor array composed of two gravimetric sensors based on zeolitic imidazolate frameworks; (ii) generating a synthetic (gas composition, sensor array response) data set pertaining to fruit ripening rooms, where CO_2, C_2H_4, and H_2O vary in concentration with anomalies in CO_2 and C_2H_4 concentrations; (iii) training and evaluating a one-class support vector machine as the anomaly detector. We show how the anomaly detection performance diminishes with the magnitude of the variance of (i) the background humidity levels and (ii) measurement noise.

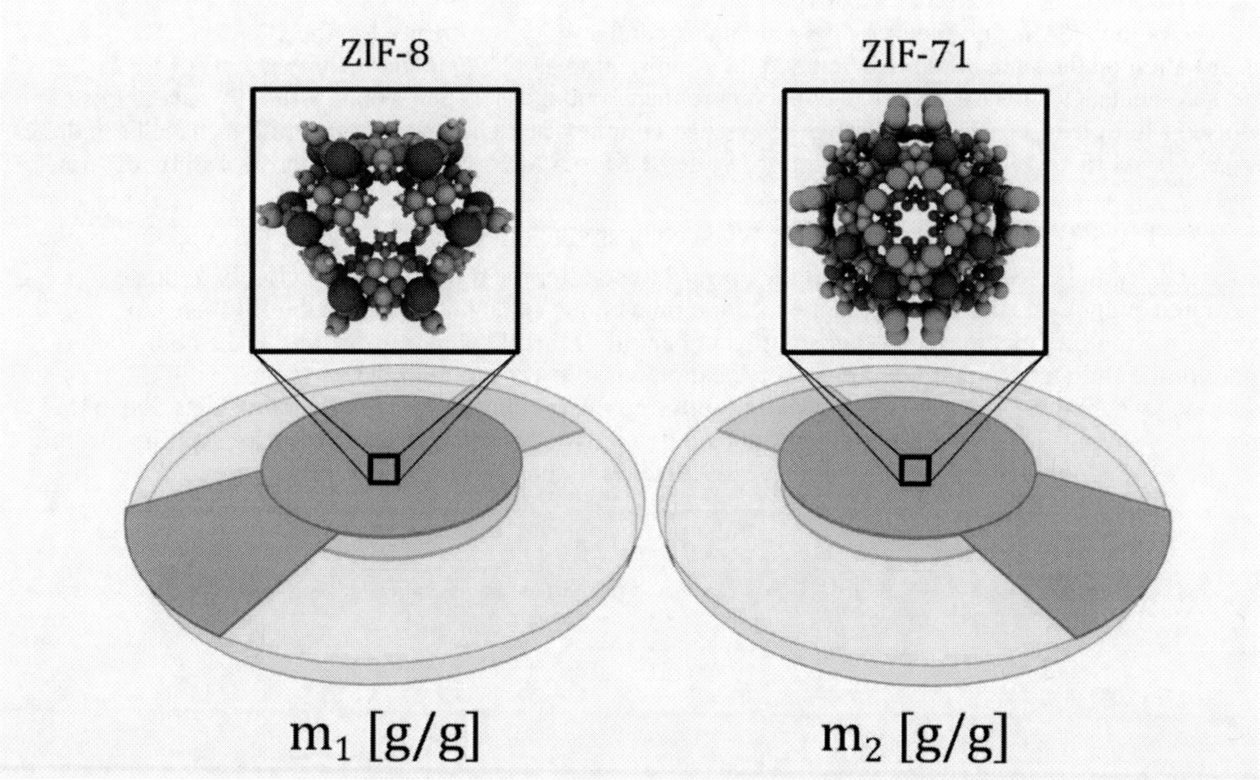

223c A Novel Sensing Technology for Online Detecting Reaction State and Fluid Mixing Using Electrical Resistance Tomography

Braden Hoff, Weiguo Xie

It is crucial to accurately determine the endpoints of reactions in the process industry, which directly influence the final product quality or downstream separation processes. At present, in most cases quality control relies on offline sample testing. However, due to this kind of offline testing being a single time point and a single location, confidence in the outcome needs to be gained through repeated testing using multiple locations. It is needed to develop a novel technology for online detecting reaction state and fluid mixing in the process industry. Electrical resistance tomography (ERT) has been applied in a wide range of scientific and engineering fields, such as geophysical study, clinical diagnosis, chemical or mineral process measurement, and process optimization. ERT is usually used to collect data on the processes in vessels and pipes. It can do online real-time measurements and without the need to draw samples from the reaction vessel that makes the data collection much easier compared to existing technologies.

In this research, ERT uses an array of 16 electrodes located equidistant in a layer around the inside of the vessel. As fluids flow past the measurement plane, conductivity is measured in the 2D plane. The information on the state of the reaction and the fluid mixing can be extracted from the data of ERT measurements. Our ERT system can collect more than 1000 datasets per second with 104 data points per dataset. Multiple test cases over a range of vessels, volumes, and chemical concentrations have been done. From 0.05 M to 4 M HCl and H_2SO_4 titrated with 0.1 M to 5 M NaOH. Figure 1 shows online real-time ERT measurements to detect the titration reaction's endpoints for (a) 0.25M HCl titrated with 1M NaOH, (b) 0.25M H_2SO_4 titrated with 1M NaOH. When compared to other methods to find the endpoints of the reactions, it can be seen that the start of the upward trend after the dip matches up closely to those measured endpoints. Additionally, more accurate time points can be obtained by the ERT system comparing to other methods. Another significant benefit of the 2D planar measuring is the ability to monitor the fluid mixing. Figure 2 reveals a location with slower mixing relative to the rest of the vessel. Knowing where these spots exist allows for monitoring the products for any potential defects due to this imperfect mixing. This novel technology can be potentially used in large reactors where visibility, testing access, or conditions restrict access to the vessel, such as higher pressure and temperatures.

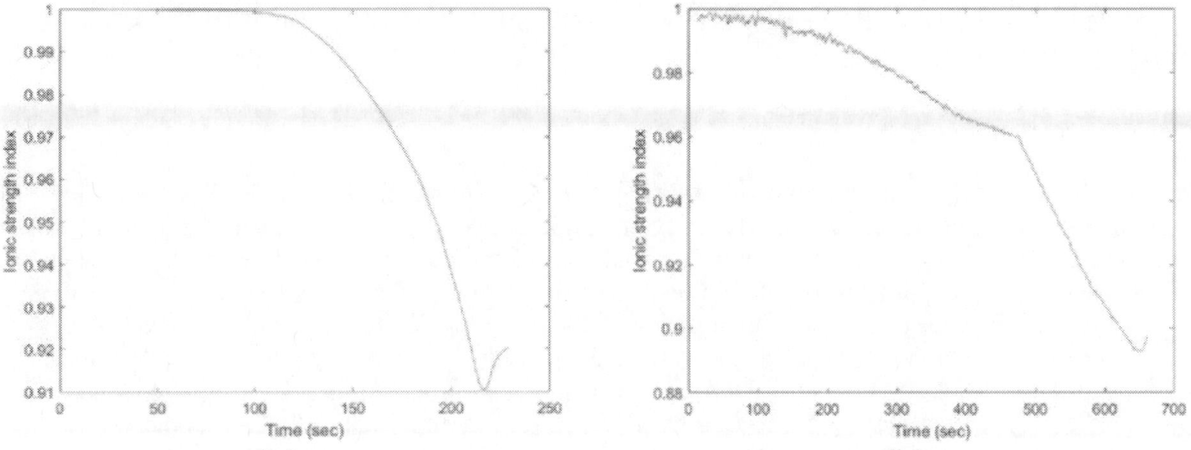

Figure 1: Online detection of the titration, (a) 0.25M HCl titrated with 1M NaOH, (b) 0.25M H_2SO_4 titrated with 1M NaOH.

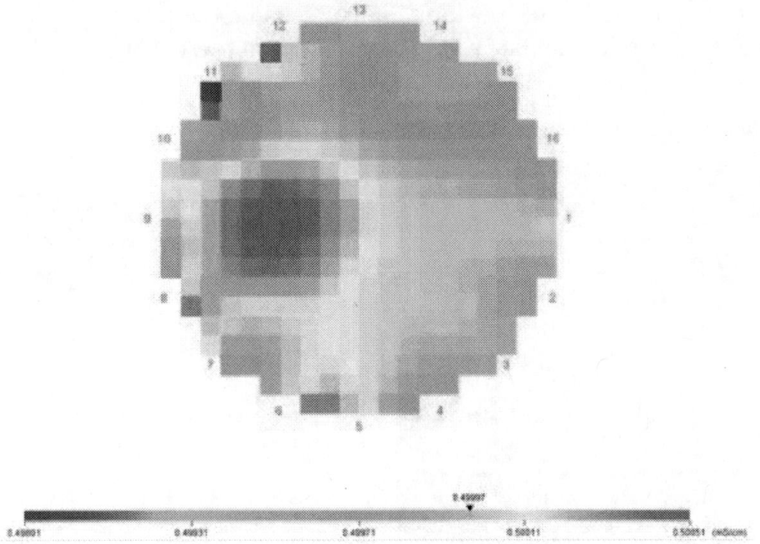

Figure 2: Spot of less mixing.

223d Ag-CNT-Silicone Based Piezoresistive Tactile Sensor with< 200 Mm Spatial Resolution for Large Area Artificial 3D-Curved Skin Deployment and Posture Stability Applications

Muhammad Haider

The soft robotics or next generation robotics, with highly compliant materials is unraveling the limitations in mimicking the human sensing abilities. Tactility and haptic sensing is very substantial in Human-Robot Interaction (HRI). Here we propose a novel piezoresistive sensor for motion detection and tactile/haptic sensing applications. The synthesis of 300 microns in diameter, highly flexible wire with piezoresistive sensing of 5% to 400% strain. Silver nanoparticles AgNPs were attached to the surface of MWCNT by repetitive chemical reduction technique. Ag-MWCNT-Silicone wires were prepared by the addition of known concentrations of Ag-MWCNT in 20 mL toluene and was sonicated for 1 hr. Homogeneous dispersion of MWCNT in silicone was induced by mixing at high temperature (70°C- 90°C) and sonication. At high temperature (80°C), relatively less viscous solution of Ag-MWCNT-Silicone was micro-molded in to the polytetrafluoroethylene (PTEE) 0.3 mm diameter tubes by high pressure syringe pump and was cured for 20 min during centrifuging at 12000 RPM. At 12000 RPM the conductive filler arranged itself on the surface of the as-synthesized wires making it highly responsive to the applied strains. The resultant wires were insulated by ≤ 50-micron silicone layer further it is analyzed for piezoresistive properties. These wires were used to form wide range of sensing devices from large area deployment tactile/haptic sensing device to human body motion detection type sensor.

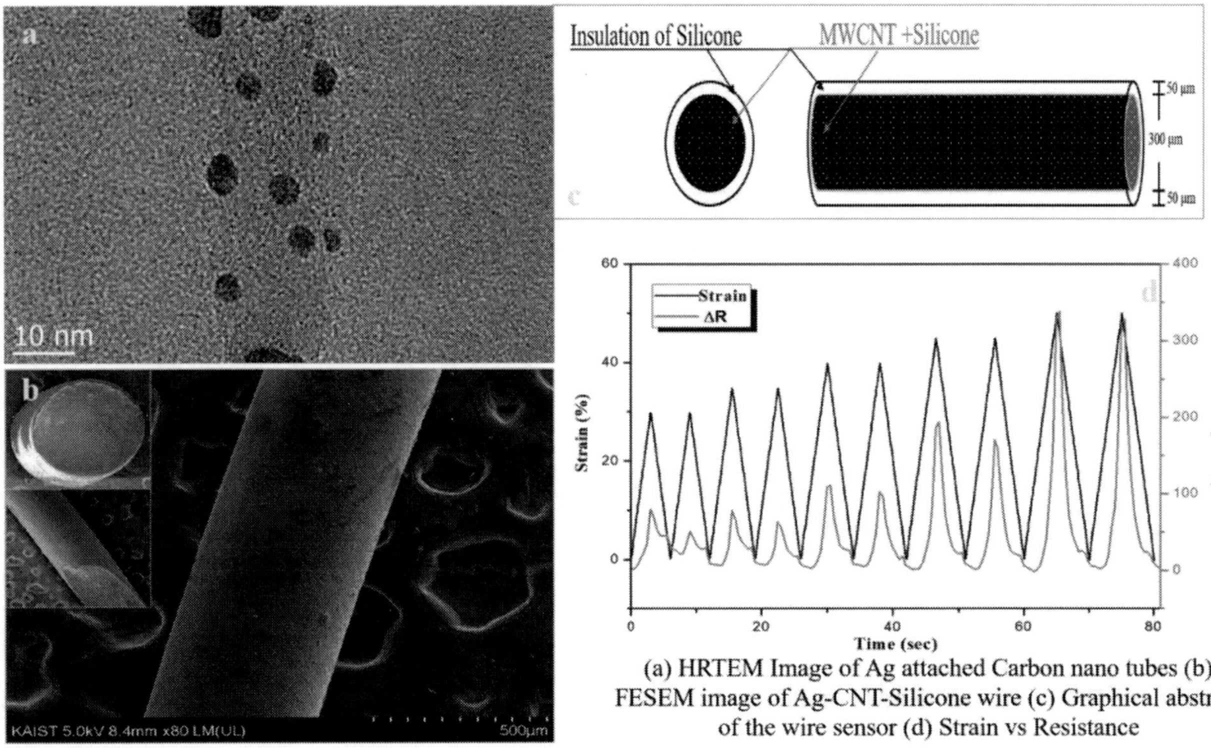

(a) HRTEM Image of Ag attached Carbon nano tubes (b) FESEM image of Ag-CNT-Silicone wire (c) Graphical abstract of the wire sensor (d) Strain vs Resistance

223f Investigation of Surface-Bound ELP As a Thermoresponsive Polymer Film

Stanley Feeney, Zahra Panahi, Eva Rose M. Balog, Jeffrey Halpern

Thermoresponsive polymer films have been investigated for their use for a variety of applications in the biomedical field, including wound dressings, tissue scaffolding, and biosensing.[1] Elastin-like polymers (ELP) are a class of genetically engineered thermoresponsive polymer that are modeled after the naturally occurring elastin protein found in the extracellular matrix. ELP are a great thermoresponsive candidate due to their ease of production and tunability; they can be expressed with high yield in *Escherichia coli* and can be genetically encoded to produce a library of variants. ELP possess a Lower Critical Solution Temperature (LCST), above which they form a condensed, undissolved state and below which they form an extended, solvated state.[2] This transition temperature has been experimentally observed many times in solution. However, surface-bound ELP have yet to be fully characterized in terms of their surface reproducibility, thermoresponsive behavior, and electrochemical drift. We use Electrochemical Impedance Spectroscopy (EIS) and quartz crystal microbalance (QCM) to monitor the thermoresponsive behavior of surface-bound ELP. Specifically, we measure the response from two ELP-modified surfaces: a thiol-bonded gold electrode surface and a covalently bonded carboxyphenyl film on glassy carbon electrodes. In addition, we compare surface modification reproducibility of the two surfaces and normalize for the inherent drift that occurs in EIS. Finally, we discuss our progress in using non-faradaic EIS to understand the thermoresponse of ELP surfaces.

Support was provided, in part, by a National Science Foundation EPSCoR award (#2119237).

[1] M.A. Ward, T.K. Georgiou, Thermoresponsive polymers for biomedical applications, Polymers (Basel). 3 (2011) 1215–1242. https://doi.org/10.3390/polym3031215.

[2] M.A. Morales, W.A. Paiva, L. Marvin, E.R.M. Balog, J.M. Halpern, Electrochemical characterization of the stimuli-response of surface-immobilized elastin-like polymers, Soft Matter. 15 (2019) 9640–9646. https://doi.org/10.1039/c9sm01681c.

223g Responsive Complex Emulsions: All-Liquid Sensors for Biomolecules and Environmental Pollutants

Suchol Savagatrup

Responsive complex emulsions provide a unique sensing platform using the tunable interactions at the liquid-liquid interfaces. In this presentation, we report our progress toward the fabrication and applications of chemical sensors based on complex emulsions for distributed real-time sensors. These dynamic, multicomponent emulsions behave as a natural sensor with reconfigurable morphologies that are extremely sensitive to the chemical environment. Specifically, the unique coupling among their chemical, physical, and optical properties can be leveraged to detect different classes of biomolecules and environmental contaminants. Specifically, our sensors detect the changes in opacity, direction of fluorescent emission, and fluorescent transmission, respectively. The selective recognitions that give rise to changes in the optical properties are governed by functionalized responsive surfactants and interfacial tensions. Furthermore, we will discuss our recent effort to detect a class of emerging contaminants, per- and polyfluoroalkyl substances (PFAS). We test the hypothesis that different behaviors at complex interfaces will give rise to the discriminating power necessary to distinguish PFAS from hydrocarbon surfactants. We anticipate that the incorporation of responsive all-liquid materials will provide a cost-effective, easily tunable, and portable alternative to laboratory-based analytical methods for on-site sensing applications.

223h On-Chip Infrared Spectroscopy with Near-Field Detection

Sean McSherry, Andrej Lenert

The miniaturization of infrared spectroscopy can provide a cheap and portable platform for molecular identification across a large application space. Infrared (IR) spectroscopy enables the identification of molecules by correlating absorption peaks to vibrational modes of atomic bonds such as C=O. Two major technical challenges associated with on-chip IR spectroscopy include i) the lack of successful integration of broadband infrared emission sources, and ii) the selectivity of the photodetector to spectrally resolve different wavelengths. To overcome both these challenges, we introduce an on-chip infrared spectrometer equipped with a set of "all-in-one" infrared sensors that both source and detect outgoing infrared radiation to an analyte. Each sensor has a unique narrowband emission/detection window, which cumulatively makes for a broadband IR detection range. A single sensor is made from a suspended, resistive calorimeter embedded in a substrate and separated from a membrane-support by ~500 nanometers (see Figure 1). Nanoscale separation distances ensure the calorimeter only sources and detects if the analyte has a vibrational mode (absorption peak) that overlaps with the narrowband, near-field emission/detection window of the calorimeter. We demonstrate a silicon nitride calorimeter that can distinguish between water and ethanol by spectrally matching the near-field detection window to distinct vibrational modes. Overall, our near-field detection technique paves the way for broadband on-chip spectroscopy with the potential to decrease the cost of an IR spectrometer to below $50.

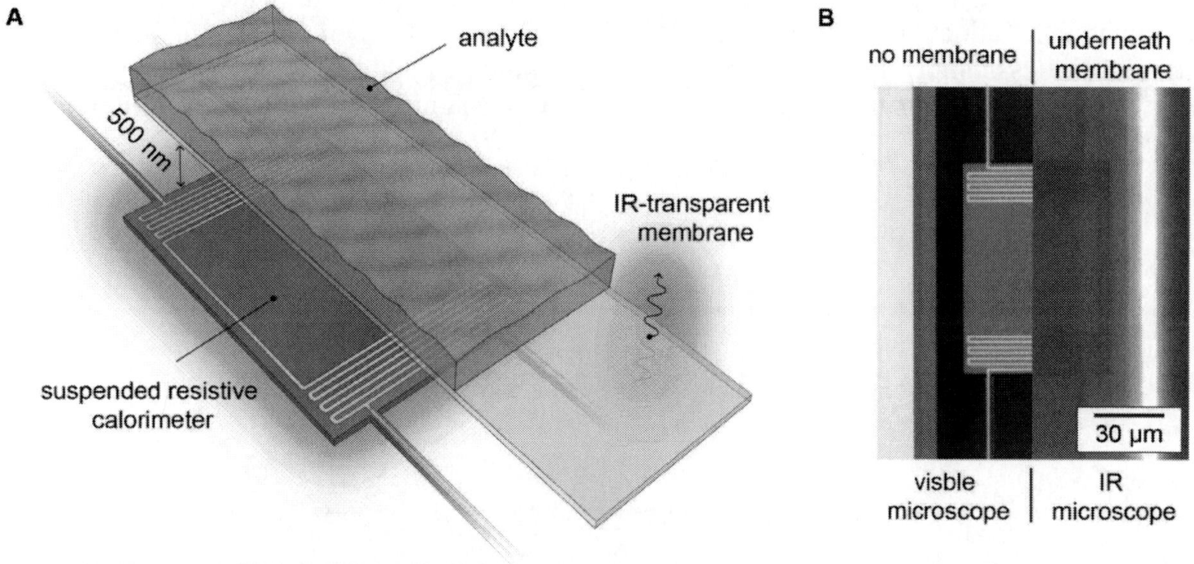

Figure 1| A. Schematic of "all-in-one" infrared sensor made from a suspended, resistive calorimeter separated from a membrane-support by ~500 nm. **B.** The left image is an optical microscope reflectance image of the suspended calorimeter and the right image is a transmission IR microscope image of the calorimeter after it is embedded underneath the membrane.

223i Possibility of Energy Harvesting Wireless Sensors Using Magnetic Phase Transition

Yasuki Kansha, Hikaru Kiyomoto, Yuka Sakai, Yuki Sato

The principles of 'Industry 4.0,' which originated in Germany, are now widely known. Following this trend, most industries are connected by cyber-physical systems (CPS), and data and energy in manufacturing technologies are exchanged and shared. Recently, many countries have interested in CPS into societies for sustainable development. CPS equips a data acquisition function, a data storage function, and a network to transfer data. To deploy CPS, it necessary to develop overall security systems, to increase their energy efficiency and to find new energy sources. In meeting the requirements for electricity for CPS, the term 'energy harvesting' is commonly used. However, the energy efficiencies of currently available energy harvesting technologies are still low. Furthermore, they often have energy mismatch with demands.

In this research, authors proposed innovative energy harvesting data acquisition sensors and investigated the possibility of these energy harvesting sensors in which data acquisition sensors were integrated with energy harvesting systems to improve energy efficiency and to prevent the energy mismatch. The targeted energy harvesting systems, which the authors have developed, equip a magnetic material and a solenoid. The magnetic phase transition of the magnetic material was taken places at the Currie temperature and electromotive forces was observed by the solenoid following the electromagnetic induction due to the magnetic flux changes in temperature, results in that this system works as wireless temperature or velocity sensors without any additional energy input. Our proposed energy harvesting sensors have the potential to contribute significantly to the development of CPS in the near future.

157b A Low-Cost and Facile Color-Changing Nanofilm for Temperature Monitoring and Recording

Md Nayeem Hasan Kashem, Wei Li

To help protect the integrity of bioreagents such as COVID-19 vaccines, a reliable temperature monitoring, and recording device is needed to prevent them wasted due to loss of potency from out-of-range temperature storage in manufacturing or transportation. Current electronic-based temperature monitoring and logging devices are costly, not directly monitoring the temperature of the bioreagents, and facing cyber insecurity with data transmission. A non-electronic-based temperature monitoring and recording sensor based on the inherent properties of smart materials can be a potential solution to overcome the limitation of current devices. Smart nanofilms can be designed to be sensitive to different external parameters such as pH [1, 2], temperature [3, 4], light [5, 6], magnetic field [7, 8], etc. However, incorporating sensitivity to external parameters in the nanofilms mostly requires sophisticated fabrication techniques. In this work, a smart layer-by-layer (LbL) nanofilm has been developed to monitor and record the temperature change of the surrounding environment by changing its inherent structure color.

Previously, a humidity-sensitive nanofilm was fabricated in our lab using the LBL technique [9]. The film showed inherent colors depending on its thickness. Moreover, it could be crosslinked by UV-light and thus designed patterns could be created on it. Under higher relative humidity (even human breath), the film absorbs moisture and could reveal the UV-crosslinked patterned structures. The film was fabricated using chitosan and carboxymethyl cellulose azido derivatives (CMC-N_3), which were deposited on a silicon wafer alternatively as per the LbL technique. With further investigation, it was found that the film could change its color when changing the surrounding temperature under a fixed relative humidity (RH). As the chitosan and CMC-N_3 are polyelectrolyte materials, they are both hydrophilic. So, the film could absorb or desorb moisture from the air and show different thicknesses at different temperatures under a certain RH. When the natural light fell on the film with a certain thickness, it went through the layers of the film to the silicon wafer, reflected, and went through the film again to the outside of it. In different layers, the wavelength of light was absorbed partially, and based on the wavelength of the reflected light a certain color can be seen. As the film showed different thicknesses at different temperatures, the wavelength of the reflected light was different, thus, the color change occurs. This type of thickness-depended color change can also be seen in nature e.g., in Chameleons. They can alter their skin color by changing the distance between the guanine (photonic) crystal layers situated in the upper layer of their skin and thus changing the wavelengths of the reflected light [10].

As shown in the attached Figure, color codes of the film at different temperatures at 40% RH have been achieved and demonstrated the film as a potential temperature sensor. With controlled humidity, this film is expected to change its color irreversibly when the surrounding temperature is above a certain level. If a serial of temperature sensors is incorporated, they can record the history of temperature for the storage samples during a designed time frame. For example, a blue-purple colored film will indicate that the storage condition has been disrupted to over 30^0C. The film is low-cost and easy to fabricate using chitosan and CMC-N_3, which are both biocompatible and environment-friendly. One unique application of this film is to monitor and record storing, shipping, and handling conditions of temperature-sensitive bioreagents (i.e. COVID-19 vaccines). Furthermore, this film is expected to be potentially used for anti-counterfeiting, smart display devices, and other optical sensors.

Reference:

1. Kharlampieva, E. and S.A. Sukhishvili, *Release of a dye from hydrogen-bonded and electrostatically assembled polymer films triggered by adsorption of a polyelectrolyte.* Langmuir, 2004. **20**(22): p. 9677-9685.
2. Déjugnat, C. and G.B. Sukhorukov, *pH-responsive properties of hollow polyelectrolyte microcapsules templated on various cores.* Langmuir, 2004. **20**(17): p. 7265-7269.

3. Nolan, C.M., M.J. Serpe, and L.A. Lyon, *Thermally modulated insulin release from microgel thin films.* Biomacromolecules, 2004. **5**(5): p. 1940-1946.
4. Serpe, M.J., et al., *Doxorubicin uptake and release from microgel thin films.* Biomacromolecules, 2005. **6**(1): p. 408-413.
5. Dai, Z., et al., *Downhill energy transfer via ordered multichromophores in light-harvesting capsules.* The Journal of Physical Chemistry B, 2002. **106**(44): p. 11501-11508.
6. Zhu, H. and M.J. McShane, *Macromolecule encapsulation in diazoresin-based hollow polyelectrolyte microcapsules.* Langmuir, 2005. **21**(1): p. 424-430.
7. Fang, M., et al., *Magnetic bio/nanoreactor with multilayer shells of glucose oxidase and inorganic nanoparticles.* Langmuir, 2002. **18**(16): p. 6338-6344.
8. Lu, Z., et al., *Magnetic switch of permeability for polyelectrolyte microcapsules embedded with Co@Au nanoparticles.* Langmuir, 2005. **21**(5): p. 2042-2050.
9. Dong, Z., et al., *Photopatternable Nanolayered Polymeric Films with Fast Tunable Color Responses Triggered by Humidity.* Advanced Functional Materials, 2019. **29**(43): p. 1904453.
10. Teyssier, J., et al., *Photonic crystals cause active colour change in chameleons.* Nature Communications, 2015. **6**(1): p. 6368.

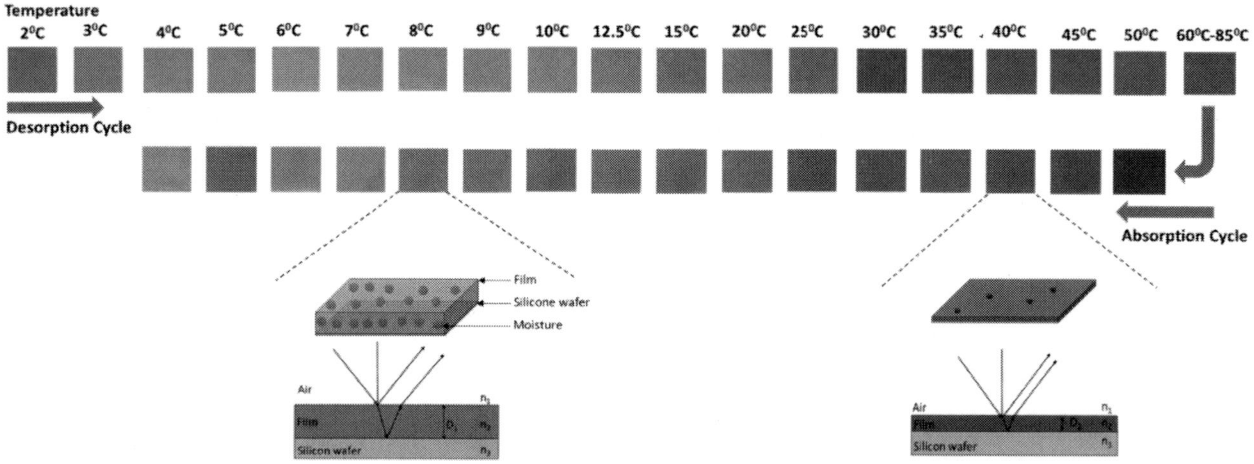

157c Using Molecular Dynamics Simulations to Model Entropic Changes upon Peptide Binding

Jazmine Torres, Ryan Peterson, Christopher Kieslich, Robert Pantazes

Peptides are promising recognition elements in many biosensing applications. Small in size and easy to produce through both chemical synthesis and biological expression pathways, peptides have many favorable features for use in biosensors. Like all thermodynamic processes, peptide binding is governed by changes in Gibbs Free Energy (ΔG). In turn, ΔG is calculated from changes in Enthalpy (ΔH) and Entropy (ΔS). An issue of current molecular mechanics force fields is the significantly limited manner in which they predict ΔS, consequently giving flawed predictions of ΔG. The motivation of this project is the development of a model that accurately predicts ΔS of peptide binding.

Molecular Dynamics (MD) is a computational method to simulate macromolecular behavior over time. In this study, we focused on small peptides of approximately 10 amino acids in two distinct states: unbound alone in solution as well as bound to a small protein domain in complex. Initial conformations of the peptides were extracted from the RSCB Protein Data Bank (PDB). Unnecessary molecules surrounding the system were removed using Chimera before being prepared using the MD software Visual Molecular Dynamics (VMD) and its QwikMD plugin. 2.5 ns MD simulations of the bound and unbound peptides were performed with explicit solvent molecules, and 100 conformational frames were generated over the course of the simulations. These frames were then clustered to identify the number of structural conformations the peptide experiences in its bound and unbound states. By clustering the repeated conformations being simulated, microstates can be quantified and applied using Boltzmann's equation, allowing for the calculation of ΔS.

This presentation will discuss our findings from an initial set of 10 diverse peptides, as well as how lessons learned from those peptides are informing a larger study with the overall goal of reliable prediction of ΔS changes from peptide binding.

157d A Capacitive Electronic Tongue Based on Cyclodextrins for Rapid Assessment of Sepsis Biomarkers

Zahra Panahi, Jeffrey Halpern

Sepsis, a life-threatening medical condition and a major cause of in-hospital deaths, contributes to more than 250,000 deaths in the United States. Early administration of antibiotics and infusion of intravenous fluids is essential to decrease in-hospital mortality rates. Development of a device for early diagnosis of sepsis, monitoring the course of the disease, and monitoring septic patients' response to the treatment is necessary for full recovery.

In this work, we employed cyclodextrins (CDs) to design an in-situ cross-reactive sepsis biosensor. CDs are oligosaccharides consisting of a hydrophobic cavity and a hydrophilic exterior. CDs are composed of six (αCD), seven (βCD), or eight (γCD) glucose units and they form host-guest inclusion complexes with various hydrophobic guest analytes. The formation and stability of inclusion complexes depend not only on the size of the cyclodextrin and guest analyte, but also the number and placement of hydrogen donors. Therefore, various cyclodextrin surfaces can be integrated to create a cross-reactive sensing array for *recognition* of a class of hydrophobic molecules.

For the first part of this project we improved the stability and reusability of our previously developed αCD biosensor by replacing gold-thiol bonds with diazonium salt chemistry on glassy carbon (GC) surface. We used polypropylene glycol (PPG) as a support for βCD and used the GC-carboxyphenyl-PPG:βCD surface for detection of a sepsis biomarker, cortisol. Our results showed that when βCD surface is introduced to the solution containing cortisol, βCD molecules leave the PPG surface to interact with cortisol in the solution. After use, the surface could be regenerated by reloading of βCD. We could successfully regenerate the GC-carboxyphenyl-PPG:βCD surface for ten times. Besides, the βCD/PPG-carboxyphenyl-GC surface was shown to be selective toward cortisol as opposed to uric acid, acetaminophen, and resveratrol. However, the selectivity of individual CD surfaces was still limited and development of cross-reactive CD sensor is necessary for clinical application.

In the second part of the project, we created five stable diazonium salt based CD biosensors with αCD, βCD, γCD, hydroxypropyl-βCD, and hydroxypropyl-γCD. We employed the developed CD biosensors for detection of various sepsis biomarkers. Linear discriminant analysis (LDA) and principal component analysis (PCA) were used to simplify the data, reduce variables, separate, and classify the data into different groups. The developed cross-reactive CD biosensor can potentially be used in urinary catheters for diagnosis and prognosis purposes.

157e Nonfluorescent CRISPR-Cas12a Biosensor By Sizing λ DNA

Noor Mohammad, Shrinivas Siddhivinayak Katkam, Qingshan Wei

CRISPR-based biosensors have shown great potential in the diagnostic field. One current limitation is that these CRISPR biosensors often require colorimetric, fluorescent, or electrochemical reporters which are expensive and/or rely on sophisticated readout devices to acquire signals. Here, we developed a simple and inexpensive CRISPR-Cas12a-based assay platform to detect short single-stranded (ss) DNA targets using double-stranded (ds) λ DNA as a novel reporter molecule. In this detection platform, the target was mixed with LbCas12a-gRNA in a microcentrifuge tube. LbCas12a became *trans*-activated, and λ DNA was then added to the reaction mixture. λ DNA was chopped into fragments and visualized in gel electrophoresis (Figure 1). The physical size change of λ DNA was quantitatively correlated to the analyte concentrations. This assay was built on the new discovery that ds λ DNA can be *trans*-cleaved by the Cas12a effector. We hypothesize that the presence of D loops and short ss loop structures along the λ DNA sequence are the keys to the massive *trans*-nuclease activity of LbCas12a toward λ DNA. In addition, a strong binding affinity between Cas12a and λ DNA was also observed, which might promote the trans-cleavage activity. Using this nonfluorescent assay, we demonstrated a limit of detection (LOD) of 220 fM for synthetic ssDNA targets without any preamplification steps, which is ~100 times more sensitive when compared with the conventional CRISPR-Cas12a assay based on fluorescent reports (e.g., fluorophore-quencher reporters). This concept of utilizing the physical size change of λ DNA unlocks the possibility of using a variety of dsDNA as nonoptical CRISPR reporters.

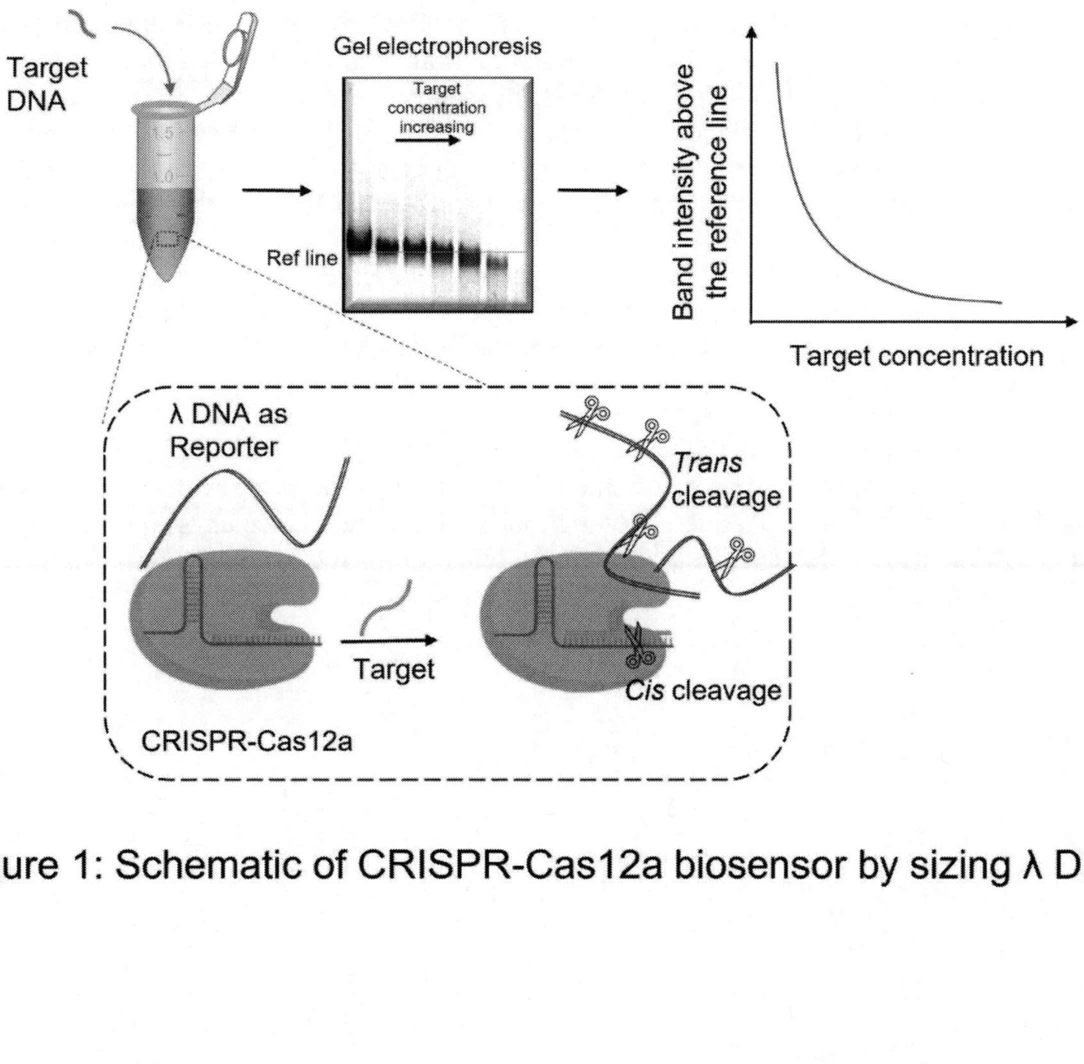

Figure 1: Schematic of CRISPR-Cas12a biosensor by sizing λ DNA

157f Point-of-Use Fiber-Based Haloacetic Acid Sensor for Water at Sub-USEPA Regulation Limits

Hadi Rouhi, Colton Duprey, Elham Ghalavand, Emily Linn, Sarah Veres, George Chen, Leigh Terry, Mark Elliott, Evan Wujcik

With the development of industries across the world, the amount of pollution released into the ecosystem has increased significantly, creating serious issues related to water pollution. The entry of halogenated compounds—including haloacetic acids (HAAs),a family of chemicals formed during the water treatment process—into treated water has been so severe that the US Environmental Protection Agency (USEPA) has limited the maximum contaminant level (MCL) in drinking water to less than 60 ppb. HAAs including five related chemicals: trichloroacetic acid, dichloroacetic acid, monochloroacetic acid, dibromoacetic acid, and monobromoacetic acid. These can enter the human body through ingestion, inhalation, and epidermal routes causing serious health problems such as colon and bladder cancer.

In this study, syndiotactic polyethylene (sPP) fiber mats electrospun by a DC electrospinning device are used to preconcentrate and detect different compositions of HAA in water samples. A well-known reaction chemistry – the Fujiwara reaction- has been used to give a red/pink chromophore under visible light absorption, which can detect combined HAA concentrations in the water samples.

The present highly sensitive and selective sensor works well when the reaction chemistry has no contact with water - a barrier to the Fujiwara reaction. Because of the detrimental effects of water on the Fujiwara reaction, the hydrophobicity of the electrospun fiber mats has been increased by increasing the roughness of the surface and spinning sPP fiber mat with a beaded fiber morphology. These fibers result in a superhydrophobic mat with a water contact angle over 150°. When water samples are exposed to temperature around 80°C, preconcentrated HAA in the vapor phase -between water sample surface and electrospun mat – becomes trapped in the liquid phase of Fujiwara reactants present on the surface of the sPP fiber mat. A yellow color indicates the ppb level of HAAs and can be analyzed with an image analysis software to measure the corresponding color intensity of the specific HAA concentration. Calibration curves obtained by these analyses can help us to accurately predict the level of HAA contaminant in complex environmental samples.

In this work, we were able to predict the concentration of HAAs in water samples down to a level of 6 ppb, which is one order of magnitude below the USEPA limit for drinking water. The advantages of this method over conventional HAA detection methods are the inexpensiveness, portability, and sensitivity of the sensor, which enables point-of-use detection of water contaminants in the simplest and quickest manner.

157g Electrochemical ELISA Biosensor for SARS-COV-2 Nucleocapsid Detection

Katherine Austin, Xingcheng Zhou, Ariel Furst

The recent COVID-19 pandemic has led to unprecedented demand for detection of the SARS-CoV-2 virus in infected individuals. The gold standard of detection includes PCR testing as well as lateral-flow tests. However, PCR testing is a costly technique due to the necessary lab instrumentation and also requires a high level of training. Rapid antigen tests are quick and inexpensive but suffer from a high number of false positives. In recent years, electrochemical techniques have gained interest within the biosensor community due to their sensitivity, relative low-cost, and selectivity. Handheld potentiostats allow the user to perform a multitude of different measurements at low cost and in non-laboratory settings. Enzyme linked immunosorbent assays (ELISAs) are a sensitive and selective technique currently used for colorimetric measurements to detect viral proteins. Due to the low cost of handheld potentiostats, ELISAs can be adapted to use electrochemical measurements for cheaper operation. We created a low-cost and highly sensitive sandwich ELISA on commercially available screen-printed carbon electrodes. By utilizing chronoamperometry we were able to detect the nucleocapsid protein (N-protein) of SARS-COV-2 via current and charge changes in the system.

157h Gold Leaf Electrochemical Biosensors for Detection of Infectious Diseases

Joshua Chaj Ulloa, Marjon Zamani, Xingcheng Zhou, Ariel Furst

Infectious diseases are significantly more prevalent in low-resource settings (LRS) due to inadequate access to preventable diagnostic testing. A major area of research is using inexpensive and simple equipment for the development of point-of-care (POC) diagnostics that can be carried out in LRS. Biosensors comprise many POC diagnostics. Specifically, electrochemical biosensors utilize biorecognition elements coupled with electrochemical transduction to enable portable, sensitive, and specific detection of the target analyte. Gold electrodes are a common substrate for electrochemical biosensors due to strong gold-thiol coupling. However, gold electrodes are too costly and their fabrication is too laborious for implementation in LRS. To make gold electrodes more broadly accessible, our lab has worked on developing gold-leaf electrodes that are affordable and do not require any specialized equipment to fabricate, making them suitable for use in LRS. We have demonstrated that these electrodes can detect human papillomavirus (HPV) from clinical samples, although this was limited to single plex disease detection. Our current work focuses on further developing the gold leaf electrodes to enable multiplexed detection. In pursuit of this goal, we have redesigned the electrodes to enable spatial multiplexing of numerous pathogens Specifically, we are using the electrodes in conjunction with CRISPR-based assays to simultaneously detect *mycobacterium tuberculosis* and SARS-CoV-2. Concurrently, we are using the electrodes to study protein-nucleic interactions with the goal of developing improved hairpin biosensors that can also be used for disease detection. Overall, we have expanded the functionality of the gold leaf electrodes so they can be used for a wide range of diagnostics in LRS.

352b Skin-Inspired Sensors for Human Body, Brain and Gut

Zhenan Bao

Skin is the body's largest organ. It is responsible for the transduction of a vast amount of information. This conformable, stretchable, self-healable and biodegradable material simultaneously collects signals from external stimuli that translate into information such as pressure, pain, and temperature. Over the past decade, we have developed materials design concepts to add skin-like functions to organic electronic materials without compromising their electronic properties. These new materials and new devices enabled development of skin-like sensors for a broad range of applications in medical devices, robotics and wearable electronics. In this talk, I will discuss our recent sensor developments to allow mapping bioelectrical signals and neurotransmitters in addition to physiological signals.

Cell-free synthesis goes electric: dual optical and electronic biosensor via direct protein integration into a supported membrane electrode

Professor Susan Daniel
Cornell University
RF Smith School of Chemical and Biomolecular Engineering

Transmembrane proteins (TMPs) are Nature's biorecognition, sensing, and transduction elements and critical components in biomimetic sensing. However, they are among the most challenging biomolecules to integrate into biosensors because they require a lipid bilayer to remain functional. Successful bilayer and TMP integration with bioelectronic devices would enable new capabilities in *in vitro* biosensing and biological actuation. To date, reconstitution of lipid bilayers onto sensing surfaces requires significant optimization of the abiotic/biotic interface. Nonetheless, current hybrid biotic-abiotic devices show great promise in biosensing, but even more so if arrays of TMPs tailored to sense specific targets could be produced. Such scale up is necessary for rapidly identifying the most sensitive protein sensors, providing a myriad of sensing elements that can monitor a variety of targets simultaneously. Our team has developed several approaches to integrate TMPs into supported lipid bilayers that functionalize the surface of organic bioelectronic devices. In this presentation, I will highlight an approach we pioneered using cell-free synthesis to integrate TMPs directly into lipid bilayers that coat the sensing surface of a planar electrode. We sense TMP properties, functions, and responses to chemical stimuli using the dual modality of electronic and optical means. The criticality of dual mode data collection is that, for transporters and ion channels, electrical means can be used label-free to read out flux of material across the membrane. Optical approaches offer the possibility to simultaneously obtain confirmation of folding with protein folding reporters, as well as obtaining orientation information. Combining all these features into one platform offers a wealth of possibilities for many biosensing and bioanalytical applications, including those that leverage TMPs for environmental sensing.

352d Polymer-Based Sensors to Monitor Indoor Air Quality and Ocular Response

Bryan Boudouris

Sensing elements continue to play a crucial role in myriad human health efforts, and devices based on functional polymers offer a means by which to usher in the next generation of advanced sensors. Here, we describe to distinct efforts with respect to this design archetype. In the first of these efforts, we combine a tailored blend of poly(ethylene oxide) (PEO) and polyethyleneimine (PEI) with microelectromechanical system (MEMS) resonant mass sensors to yield low-cost, low-power consumption carbon dioxide sensors. In this materials system, the PEI serves as an effective sorbent for carbon dioxide uptake while the PEO moiety aids in guiding the structure of the polymer blend. By controlling the polymer formulation and the processing procedure used to deposit these materials from solution, we control the micro- and nanoscale structure of the final polymer thin films. This control over the local structure allows for sensors that are highly specific to carbon dioxide relative to multiple potential interferent gases (e.g., carbon monoxide and ethanol). Moreover, the PEO-PEI sensors show clear response signals across a broad range of carbon dioxide concentrations (i.e., from 100 ppm to 10,000 ppm), and this is true in air-only background environments and in cases where a residual background of carbon dioxide is present and across a range of relative humidity values. Thus, these sensors can be utilized as proxies for room occupancy in buildings, which is becoming of increasing importance given the global pandemic and the need to minimize energy usage in the built environment.

In the second effort, we demonstrate the design of all-printed stretchable corneal sensor built on commercially-available disposable soft contact lenses that can intimately and non-invasively interface with the corneal surface of human eyes. This is critical as electroretinogram (ERG) examinations serve as routine clinical procedures in ophthalmology for the diagnosis and management of many ocular diseases. However, the rigid form factor of current corneal sensors produces a mismatch with the soft, curvilinear, and exceptionally-sensitive human cornea, which typically requires the use of topical anesthesia and a speculum for pain management and safety. On the other hand, our poly(3,4-ethylenedioxythiophene)-based (PEDOT-based) corneal sensor is integrated with soft contact lenses via an electrochemical anchoring mechanism in a seamless manner that ensures its mechanical and chemical reliability. Thus, the resulting device enables the high-fidelity recording of full-field ERG signals in human eyes without the need of topical anesthesia or a speculum. The device, superior to clinical standards in terms of signal quality and comfortability, is expected to address unmet clinical needs in the field of ocular electrodiagnosis. Moreover, we modified the general design of the sensor such that a new iteration serves as a means by which to measure interocular pressure. We did this because continuous monitoring of intraocular pressure remains a challenge in glaucoma care, especially during sleep. Specifically, we introduced an intrinsically soft and stretchable ocular tonometer built on a commercial overnight-wearable soft contact lens. The resulting ocular tonometer, or contact lens sensor, fits seamlessly across different corneal curvatures and thicknesses in human eyes and is capable of wirelessly capturing absolute intraocular pressure levels without iterative calibrations. We validated the biocompatibility, wireless sensing capability, on-eye safety, and 24-hour monitoring reliability of the contact lens sensor in human corneal cells *in vitro*, pig eyes *ex vivo*, and rabbit and dog eyes *in vivo*, respectively. We also showed the continuous monitoring of ambulatory intraocular pressure in human eyes under normal and ocular hypertension conditions to confirm the measurement accuracy, within-subject repeatability, and user comfort of the contact lens sensor beyond current state-of-the-art wearable ocular tonometers. We envision that the new contact lens sensor become the gold standard in managing glaucoma and ocular hypertension, especially as telehealth treatments continue to rise. In this way, we aim to have this talk describe two key areas where polymeric materials can improve sensors that will allow for enhanced human health.

352e Functional Biosensors for Human and Environmental Health

Ariel Furst

Electrochemistry is an exceptionally powerful tool to monitor chemical and biological interactions. By harnessing the inherent activity and function of biomolecules, we have significantly improved the selectivity and sensitivity of electrochemical biosensors. Specifically, by combining electrochemistry with synthetic biology, we have developed a platform to detect biologically-active small-molecule environmental pollutants at sub-micromolar levels. Similarly, by harnessing the inherent metabolic activity of pathogens, we can detect bacteria that cause food- and water-borne illnesses. By combining improved chemistries for biomolecule modification with unique signal amplification strategies, we have successfully detected targets from extremely complex solutions and directly incorporated remediation with detection.

AUTHOR INDEX